FARMING IN THE NORTH

by

Hunter Adair

HUTTON PRESS

2003

Published by
The Hutton Press Ltd.,
130 Canada Drive, Cherry Burton,
Beverley, East Yorkshire HU17 7SB

Printed and bound by
Intellectual Print
01482 222778

ISBN 1 902709 23 3

CONTENTS

INTRODUCTION

The north is a very mixed agricultural area, and some of the best cattle and sheep in the country are produced in this region of the United Kingdom. There is also a vast difference in the size and geography of the farms, and although these may be only a few miles apart, they can be very different because of the climate.

Some farms in the Dales produce lambs and young beef cattle, which are then sold on to the richer pastures in the south, where they are finished off for the market. There are now very few small dairy farms in the Dales producing milk from the numbers nearly thirty years ago. Most farmsteads then kept a few dairy cows, and many families were brought up on smallholdings, which have now sadly disappeared.

Some farms in the north are also well known for growing high yielding quality cereal crops, and quite a large amount of the cereals are sold for export. The climate in this region is very changeable, although the winters are not nearly as hard as they used to be. When you get a cold wet winter and spring, it has a great effect on farming, and the quality and quantity of the products produced.

The north of England is well known for its richness of wildlife, although over the years the development of agriculture has led to the disappearance of the corncrake from the countryside. Things are now changing, and the corncrake has been heard again, which is a very good sign.

This book is about a cross-section of farming and wildlife in the north over a number of years, from Yorkshire to Scotland. There is a range of photographs and drawings of farming and wildlife to support the text. Most of these are by the author.

CHAPTER 1

THE AYRSHIRE COW

There are many breeds of both dairy and beef cattle in Britain. Some farmers specialise in certain breeds of cattle to a very high standard for various reasons. It is sometimes because a particular type of animal suits their farm, or they may just prefer a specific type of breed, and specialise in them.

The Ayrshire dairy cow, for instance, was developed in the county of Ayr, on the south west coast of Scotland. It was bred from the Shorthorn, the Channel Island and the Highland cattle, from about 1750. The breed has been very popular in Tynedale. The first variety of the cattle were called the Dunlop breed after the name of Mr. John Dunlop, who was probably the greatest of the early breeders. However, in 1814, they became officially known as Ayrshires, and in 1877 the Ayrshire Herd Book Society was formed.

Before 1800, many of the cattle in Ayr were black from the old black Celtic cattle, although browns and mottled colours were beginning to be bred into the cattle. The colour of the Ayrshire cattle today varies from reds or browns, and many of them have mottled brown spots with a lot of white. Some of the cattle are darkish brown in colour, with a few patches of white.

The Ayrshire cow is a hardy animal, as it is developed in the high rainfall area of the west coast. The breed is comparatively small, and the neat females have level top lines and tidy udders, and wear well. This type have won numerous prizes over the years at agricultural shows for their perfect dairy conformation. The quantity of the milk they produce is good in relation to the size of the cow. The butterfat content averages about 4%, and the fat globules are small so that they rise to the top slowly. This is an advantage for cheese making, and so Ayrshire milk is ideal for this purpose.

The Ayrshire breed has spread to a number of countries throughout the world, such as Canada, New Zealand and Ireland to name a few, but they are mainly a Scottish breed of dairy cow, and are most popular in Scotland. However, in 1979, a herd of eighty pedigree Ayrshire cows were flown out to Kuwait, on the Persian Gulf, one of the richest countries in the world. The herd was worth £50,000, and were sent on a year's trial. The Ayrshires are renowned for their adaptability to extreme climates, and the reason they were selected for the Persian Gulf was because it was thought that the climate of hot and cold would suit the breed. I did hear from an Ayrshire breeder in 1980 that the Ayrshires were doing alright in Kuwait, but I have not heard any more since then.

The only disadvantage with the breed that I can see, are the bull calves. Because they are fine boned, they take much longer to fatten for beef, unless the cows are crossed with a beef bull. The bull calves from the Ayrshire cows can be fattened alright, but it takes a long time. I can remember when we had about forty Ayrshire dairy cows in Scotland, and sometimes about ten or a dozen pure bred bull calves were kept to fatten for beef. We did eventually get them ready, but it took about three years. The bull calves were hand fed with turnips and crushed beans and oats throughout the winter, as well as good hay.

The temperament of the Ayrshire breed is about as varied as their colour. Some cows would wait until you were putting the milking teat cups onto the last teat, and then they would lash out at you, but if you kept tight against the cow there was less chance of them kicking you. Sometimes when the cow had finished milking and you went in beside her to take the milking units off her teats, the cow would slightly turn her head and wait until you had just pulled the milking units off her, and then she would kick at you with some force - and she meant it! The closer you kept against a kicking cow, the less chance she had at getting you, but if you stood back from her and you were within her range, she would have you, and she would really give you a hard clout with her feet. The majority of the Ayrshire cows were very placid, and you could work amongst them without any trouble, but when you got a nasty one, even if she was a good milker, she was better sent down the road for everybody's safety. When you were buying milking cows at the cattle markets, no farmer would sell his best, quietest cows, unless he had to. It was normally the dairy cows with faults that were being sold; poor milkers, kickers, and cows that were difficult to get in calf.

There are two incidents with Ayrshire cows and bulls which I will never forget. The first was when I came down to work in England with the Milk Marketing Board in 1958 at their cattle breeding centre near Penrith. I was helping the head cowman to look after and feed the pedigree bulls. There were about twenty bulls at the centre, all pedigree dairy and beef bulls. The bulls were all tied up in the one sided cowshed, and there was a feed passage in front of them where we used to walk up and down to give food to them. The bulls were used for the A.I. (Artificial Insemination) service. In front of the bulls there were bars where they could put a horn through the bar and get you if you weren't watching them. Some of them were very nasty. Our instruction was not to hit the bulls with anything, as they would remember. We had this particular very dark brown Ayrshire bull which was very nasty, and kept trying to hook us with its horn as we passed by in the feed passage. The bulls were taken out once a day for some exercise in the enclosed farmyard where the hay shed stood. There were two staff with a bull when it was getting exercised. One of the staff had a long wooden pole with a hook on the end which was attached to the bull's ring in its nose. The other had a long rope which was also attached to the bull's ring, and the rope came through between the animal's legs, so he walked behind it just for safety reasons. There was also an exercising machine where some bulls were hooked on to by the ring on their nose, and they could walk round and round for about two hours or so. This was sited in the field just beside the farm buildings. There were four staff working with the bulls.

One morning the head stockman and I had this nasty Ayrshire bull out in the yard on a pole and rope. The bull was snorting and grunting at us, when suddenly the pole in the bull's nose came off. I was behind the bull with the rope still attached to its ring. The bull turned on me, and was walking slowly to me with its head down, snorting and grunting, and kicking the soil up with its front leg. The head stockman lay flat on the ground until the bull was nearly over the top of him, still grunting and snorting. Then the stockman put his hand up and caught the bull by the ring and stood up. He then led the bull back into the byre. The other two stockmen had joined me, and watched this act of courage from their boss. This is why he was the head stockman!

The other incident I will never forget about was with an Ayrshire cow. My father-in-law bought this dairy cow at the local market here in Hexham one Friday in December 1970. When the cow arrived at our farm in the late afternoon with the cattle wagon, it was beginning to get dark, and it was a horrible wet afternoon. When the cow was let out of the wagon into the farmyard, it looked around and made a bee line for the field gate in the corner of the yard, and jumped straight over this five foot gate into the front field. I quickly followed the cow down our front field. The cow then jumped into the next field, then the next one, while I was hot on its tail. By then it was nearly dark, and it was only the lights from Hexham that provided me with enough light just to see the cow as it charged on.

On the west side of Hexham there is a large reservoir which is about twenty or thirty foot deep. I was flabbergasted as I could see the cow heading straight for it! I kept close to the cow, then I saw it jump straight into the reservoir, swimming away out into the middle of it. As I approached the reservoir, I, too, jumped into the cold water, and swam away out around the cow, and headed it back to the bank side. When I got the cow and myself out of the icy cold water, we were by now about a mile from the farm, and the cow was now heading for Hexham town. As we got into the built up area in Hexham I managed to get her into a private garden. By then I was joined by my father-in-law and our farm worker. We managed to get the distressed cow shut up in a wood shed. I went back home and got her some hay. When I opened the wood shed door, the cow charged at me. I was aware that this may happen, and just managed to fling the hay into the shed and shut the door. The next morning when we went for the cow, she was as quiet as a lamb, and she turned out to be a very good milking cow.

What a mess the cow had made of this private garden. I spent half a day filling the cow hoof marks with sand. What an experience with myself and an Ayrshire cow swimming in the reservoir at the same time!

***

A DAIRY FARMING STORY WHICH IS TRUE

I think that this story is hilarious. The Electricity Board were laying a cable underground along the side of a farmer's field where his dairy cows were grazing. The Electricity Board put a fence up between the cows and the men laying the cable. The cows were so intrigued with the men laying the cable, that they stood watching them all day, and forgot to eat the grass. When the cows came in to be milked that afternoon, they had little milk to give. The farmer then decided to claim for the loss of milk from the Electricity Board!

Wife Kathleen with suckler cows and calves.

John Huddleston, an Ayrshire breeder in the north.

A breeder of Ayrshire pedigree dairy cows in his cowshed, which holds two hundred cows.

A typical double-sided cowshed, where the cows are chained up by the neck for the winter.

A Danish Red cow imported from Denmark in 1961 by Mr. Colin Weightman from Low Shilford Farm at Stocksfield.

CHAPTER 2

THE BRITISH FRIESIAN

The Friesian dairy cow has been very popular with many farmers in the whole country. In the eighteenth century there were a lot of black and white cattle imported from Holland to Britain. Some of these animals produced plenty of milk, and many of the cows were bought by the urban and city dairies to supply milk for the expanding population in the towns and cities.

A lot of these black and white Dutch cattle were crossed with local breeds of cattle, but no real improvement was made with the stock until the second half of the nineteenth century. At this point, individual breeders of the Friesian cattle in some parts of the country started keeping records of the milk yields from each cow. From this documentation the best animals could be selected and used to breed the next generation of Friesians.

The British Friesian Cattle Society was founded in 1909, and some early members soon sponsored a number of black and white Dutch Friesians to be imported from Holland. These cattle had a great effect on the new breed of black and white Friesians being bred in this country, as many bulls being bred from the Dutch cattle became herd sires. The outcome is that this variety is now a well-known breed of dairy cow in this country, yet it also contributes greatly to the beef production in Britain, as the Friesian crosses well with our native and imported breeds of cattle which produce top quality beef for the market.

The British Friesian breed is black and white, and there is probably more black in the stock than white. The black is usually in the lower parts of the dairy cows, in the legs, the udders, the thighs and the bottom half of the tails. They are slightly bigger than the other dairy breeds such as the Ayrshires and Shorthorns, and are bigger boned. There are some top quality breeders of this breed throughout this country and they have produced some of the best quality Friesians cows and bulls in the world, and have become well-known breeders. An advantage that this breed has over the other dairy cattle is due to the quality of the calves. The bull calves fatten much quicker than those from our other native dairy breeds. Any Friesian heifer that turns out to be a duff milker could be crossed with an Aberdeen Angus, Galloway or a Hereford bull, and the cross with these beef bulls produces some top quality beef. In the 1970's, beef bulls started to be imported from Europe, one variety being the Limousin bull from France, which is a brown/red coloured variety, and is almost square in looks. The cross between the Friesian cows and the Limousin bulls produces some good top quality beef cattle. The pure bred Limousin also make good beef cattle, and many farmers in this country now breed pedigree herds. Although I like the Limousin cattle conformation, I very much like the

Friesian cross Hereford bullocks with the white face, but because of the imported continental bulls, there are less and less of these cross breeds about.

I normally buy in my bullocks at about the end of April and keep them all summer, then sell them in November hoping to make a bit of profit. The bullocks I buy in are just over twelve months old. They are put onto good grass for a few months, and I normally give them a bit of hand food a few weeks before I sell them, cow cake mixed with barley.

The British Friesian is also very well known as a dairy breed, and became the most dominant dairy breed in the United Kingdom in the 1970's and 1980's. About 90% of the milk produced in the United Kingdom then came from this breed. Their production was greater than the other breeds, such as the Ayrshires, Shorthorns, and the Jersey dairy cows. The dairy farmers at the time were paid by the quantity of milk they produced, and not the quality. The milk produced by the Friesian cows was slightly lower in butterfat than the other dairy breeds. In the 1970's there were about 50,000 dairy farmers in the United Kingdom, and the milk from the farms was all sold through the Milk Marketing Board. The Board then sold this to all the different bottling dairies, cheese making, and butter making factories throughout the country.

As the Friesian dairy breed was making progress, it was interesting to hear some of the comments by the Dales dairy farmers about this black and white cow. I sometimes heard the comment that some of the dairy farmers only kept one or two Friesian cows at the top of the byre, and the milk from these cows was solely used for washing down the byre. They thought the Friesian cows' milk was like water. (I had many a giggle at the farmers' comments).

The Milk Marketing Board for England and Wales, which started in 1933, did a good job for the dairy farmers until it was disbanded in the 1990's. It put stability into the industry, and the dairy farmers knew that they would get paid for their milk every month. The Milk Board wasn't perfect by any means. I worked for thirty years with the dairy farmers in the north, and I knew all the good and bad points of the Board. I tried very hard, and was successful at changing some of their faults over the years. They say you never miss the water until the well runs dry. Well, when the Milk Marketing Board was dissolved, the dairy companies moved in to take control of the dairy industry. The dairy farmers soon felt how much they missed the Milk Board. The dairy industry is now run by companies such as the Dairy Farmers of Britain, and First Milk, who are doing a good job for the farmers, and they cover the whole country.

On the 31st March 2001, there were about 20,000 dairy herds in England and Wales, and about 7,000 in Scotland. Since the last outbreak of Foot and Mouth Disease, these figures have been drastically reduced.

A Friesian cow being washed at a local agricultural show.

Members of the Young Northumberland Holstein Friesian Breeders' Club, having a calf show at Hexham Cattle Market.

A typical black and white British Friesian dairy cow. Many Friesian cows are now crossed with the Canadian Holstein dairy cattle, which produces a bigger animal that gives more milk.

Farmers in Northumberland looking over a young Friesian bull which was going to the Perth Bull in Scotland.

Showing a Friesian bull at Allendale Show.

Dairy cows wandering their way into the farm buildings to be milked.

An evening party was held at Maxwell's of North Doddington Farm, near Wooler in 1981, to celebrate the opening of their new cow cubicle building and milking parlour.

The Winterburn family, looking over their Friesian herd of dairy cows in the collecting yard of their dairy farm near Newcastle.

Feeding a pen of cattle in a modern farm building.

CHAPTER 3

THE SHORTHORN BREED

When I first came down to work in England in the 1950's with the Milk Marketing Board, I noticed a sturdy strain of cattle particularly in the Dales. This was the Shorthorn breed. Most of the small farms in the Yorkshire and Durham Dales had a mixture of animals at the time which were either Shorthorns, or an Ayrshire type cross of animal. I was quite impressed with the former breed, which I had come across beforehand in Scotland, but not in any great numbers, and north of the border they were usually crossbred with the Ayrshire stock.

The shorthorn cow is an old established breed of animal, and goes back to the 1580's when short horned cattle were found in the Yorkshire Dales. These cattle were probably descended from a mixture of the old black Celtic, the red Anglo-Saxon, and the broken coloured Dutch cattle.

In the north east of England many breeders were involved in trying to improve their cattle, and this was also happening in Scotland. Some progress was made in breeding the Shorthorn strain between 1730 and 1780. Then two farmers called Charles and Robert Colling became involved. These two brothers, living in County Durham, set about developing their animals by a selective breeding programme. They both became famous throughout Great Britain for breeding two specific animals, which contributed a lot to the development of the species. One animal was known as the "Durham Ox", a steer that was shown many times, and at five years old weighed 1,371 kg. (3,024 lbs.). Robert Colling reared a heifer that was a twin to a bull, and couldn't breed. This animal was fattened for show purposes, and became known as "The White Heifer that Travelled".

In 1790, Thomas Booth, a farmer from Yorkshire, purchased his foundation stock, and produced a beef strain of the Shorthorn. The Beef Shorthorn breed of Scotland, was developed from animals imported to Scotland in 1837 from the Booth herd, although there was a type of black Shorthorn oxen in Scotland before this date.

At the first General Show in Scotland, held on Thursday 26th December 1822 in an enclosure at the back of Queensberry House in the Canongate in Edinburgh, the prizes were advertised for each of the four most approved breeds of black cattle:-

1. 1st prize of Ten Guineas (£10 50p.) for a pair of oxen of the Shorthorn breed exceeding four years old.
2. 1st prize of Ten Guineas (£10 50p.) for a pair of Aberdeenshire breed not under three years old.

3. 1st prize of Ten Guineas (£10 50p.) for a pair of West Highland breed not under four years old.
4. 1st prize of Ten Guineas (£10 50p.) for a pair of Angus, Fife or Galloway breeds or any other breed not under three years old.

There were also prizes for the best and second best Ox of any breed and age showing most fat and weight.

This sturdy type of black Shorthorn animal had been in Scotland long before the breed was well developed, and the name Shorthorn was attached to this Ox type black animal at this point.

It was recorded that the arrangements for putting up and classing the different varieties of stock for the show was made by Bailie Gordon of the Canongate, His Majesty's carpenter, who was professionally employed for the job. After the competition on the Thursday, Bailie Gordon gave the general public the opportunity to see the prize animals on the Friday and Saturday. The show attracted a good deal of interest. The money collected from the general public came to £51 10 shillings (£51 50p.). The directors of the Society had agreed a sum of Seventy Five Guineas (£78 75p.) to pay expenses for the first show in 1822.

Another breeder of the Shorthorn cattle in 1837 was a farmer called Thomas Bates, who concentrated on the milking qualities of the cows, rather than for fattening for the beef market. The result is that quite a lot of the dairy cows of this variety today are strains from the Thomas Bates breed.

The first Shorthorn cattle were black in colour from the old Celtic breed, but as they developed, the broken brown, reds, and white colours of the Dutch cattle were bred into them. Today the Shorthorns vary a great deal in colour, from a bright red, to a pure white, with a mixture of roans where the two colours blend.

In the early 1960's, the quality of the Shorthorn cattle in the Dales was not very good, as many farmers kept their own bulls, and some of the cattle were being interbred. When the Milk Marketing Board started bringing their top quality pedigree Shorthorn bulls into the Dales with their Artificial Insemination service, the quality of the cattle in this area greatly improved in just a few years. By the middle of the 1960's I had seen a vast improvement in the quality of these cattle, which was very heartening.

The Shorthorn cattle were once very popular in Great Britain as dual purpose cattle, but with the competition from beef and dairy breeds from abroad, they have sadly declined at the present time.

The Shorthorn bull, Burndale Ringleader, which won the Royal Show at Newcastle in 1956.

A Shorthorn cow at Allendale Show in 1959.

A farmer and his son from Weardale in County Durham, with their Shorthorn dairy cows.

LIMOUSINS AND MILK

Young Mark Henderson from Burn Tongues and Finney Hill Green, with a young Limousin bullock, which won the Hexham Suckler Sale in October 1993.

Some of the Henderson Partners, including Derek, young Kevin, John and Edward, from Burn Tongues and Finney Hill Green, with some of their pedigree Limousin beef cattle. They also breed pedigree Leicester and Suffolk sheep.

A Limousin bull, used on many farms to cross breed with our native cattle.

Sir Simon Gourlay, left, chatting to two farmers during his visit to the Northern Region in 1988.

The late Mr. John Woodcock on the right, talking to two farmers. John was the National Farmers' Union secretary in Northumberland for many years.

Sir Stephen Roberts, centre, Chairman of the Milk Marketing Board for England and Wales, chatting to two farmers whilst on a private visit north in 1981.

Mrs. Stonehouse from Westwood Farm near Wylam in Northumberland, inspecting a new carton filling machine on Robert Graham's dairy farm in Wylam.

Dairy farmers from various parts of the north, looking at some flavoured bottled milk during a visit to a creamery at Newcastle in the 1980's.

Dairy farmers from Tynedale visiting a butter making creamery at Carlisle, to see how butter was made and packed for the market.

Farmers in the north at a creamery in Newcastle in 1990.

Princess Anne is pictured discussing the making of cheese with Bill Lockhart, General Manager of Express Dairies at Appleby Creamery in Cumberland, during her visit to the creamery on 1st July, 1985.

Milk being collected from a plastic emergency container from a farm at Allendale during a heavy snow storm in the 1970's.

Farmers from Northumberland, meeting in a pub near Newcastle-upon-Tyne to discuss the state of the dairy industry.

CHAPTER 4

THE CATTLE DROVERS

When the toll roads were being laid in this country during the eighteenth century before any main public highways were built, and before the establishment of the railways, cattle drovers were the most important long-distance travellers in the land. Some people paved paths for the local needs as they arose, or they travelled on the old trade routes, many of which were ridgeway paths dating back to pre-historic times, and which are still used today by walkers.

Some of the tracks were used by lines of heavily burdened pack horses carrying essential goods, such as salt, wool from the farms, and lead and tin from the mines. For many centuries the economic survival of many towns and villages depended upon the price the drovers were able to get for their cattle.

In 1727 it was estimated that forty thousand Scottish cattle were bought by the graziers in the south each year. These Scots runts, as they were called, came out of the cold and barren hills and mountains of their native country, and they thrived very well on the rich pastures of the south. In fact the cattle flourished so well, and the beef tasted so delicious that the English preferred the Scots beef to that from their own cattle. It was the roast beef of the English that kept the cattle drovers from Scotland in business. The Scottish cattle drovers never ate such rich pickings as roast beef, their staple diet being a handful or two of oatmeal and a few onions with a ram's horn filled with whisky, which was topped up from time to time. When they ate flesh it would either be pork, or mutton, although this was only on rare occasions.

There were hundreds of drove routes throughout Britain, which were regularly used. Many of them can still be traced today, and are in some of the wildest and most beautiful parts of the country. They make good drives or walks.

The cattle from the fairs or markets in Falkirk in Scotland came over into England at Carlisle, where they were joined by the cattle brought across from Ireland to Portpatrick, and also by the cattle from Galloway. Some cattle drovers, also from Falkirk, came south east, beginning their journey by crossing the Pentland Hills to Peebles, and over the pass known as the Cauldstane Slap. They then travelled right down through Northumberland to the River Tyne, then followed the river into County Durham, tracing the pre-historic route over the Hambleton Hills between Durham and York. Other drovers would cut across to Alwinton, and journey down through Elsdon to Newcastle. Many Scottish cattle, also moving into the north of England, were driven along the course of the Roman road, which ran from Hadrian's Wall, over Becastle Fells, and on to Kirkby Thorne, near Appleby. Between Alston and

Kirkland, the drove road ran over Melmerby Fell.

All the main drove roads were crossed by green lanes which were used for driving cattle a short distance to local markets. To ensure the enforcement of the vagrancy laws, each drover had to be licensed. He had to prove that he was over thirty years of age, had a house, and was a married man.

Despite the drovers' wandering way of life, and various wild tales recounted about some of them, they were a solid bunch of worthies, and were trusted by their neighbours to ensure the economic survival of the community by carrying money and important documents to the south, as well as driving the cattle. In return, the cattle drovers brought home goods and news of the outside world to the remote farmsteads. They were the reporters of the time, and the farmers' wives were eager to ask the drovers about the latest fashions.

There are many local place names which indicate where the drovers stopped, or passed by. The word drove, or drift, is used to designate a lane, for example Halfpenny Lane. Inns called "The Drovers", or "The Black Ox", are also common in the areas used by drovers and their cattle.

There are many tales and stories about the drovers from Scotland, and one story has always amused me very much. Some drovers had very good dogs which were a mixture of collies, spaniels, rough haired, terriers and working type corgis. The drovers' dogs were like a mixture of liquorice allsorts, but they were a main part of the drovers' trade. The drovers on the east side of the country drove cattle from Falkirk, down through Berwick, Newcastle and Durham, heading for Doncaster where the cattle were sold. When some of the drovers were returning home, they would get the boat back up the east coast to Falkirk, and let their dogs find their own way home. A little girl from Berwick was one day playing outside her house when two dogs went past. She asked her father if someone had lost their dogs. "No, no," was her father's reply, "these are the drovers' dogs coming home from Doncaster".

***

Here is a story about a collie dog. We had too many collie dogs at home, and one bitch in particular, which followed at my father's heels everywhere he went. At the cattle market one day in Ayr, my father met a farmer from Maybole in Ayrshire who said he was looking for a good working collie dog. My father said that he had a collie bitch which was about five years old that we would sell him. A deal was done, and during the next week the farmer from Maybole came and collected, paid my father for the dog, and off he went. Maybole is about twenty five miles from where we lived. One afternoon about three weeks later, the collie bitch came staggering into the farm yard in a terrible state. She was as thin as a rake, and her coat was shaggy. The collie had found her way home from Maybole. My father said that he would never again sell another collie dog as far away from home!

The Drove Road at Muggleswick, by the Old Smithy.

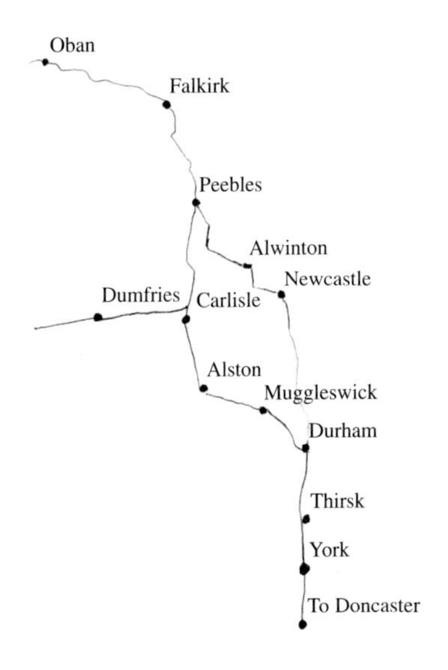

One of the main Drove Routes from Falkirk to Doncaster

CHAPTER 5

SHEPHERDS AND SHEEP

Shepherds are people who are dedicated to the craft of breeding, and the general management of sheep. To most of us, one sheep looks just like another, but to the shepherd every sheep has its own characteristics, and the shepherd knows every single one. Looking after flocks of sheep out on the hills or on lowland farms can be quite a lonely job, as quite a lot of the shepherd's time is spent alone with the sheep. There are more females now taking on the job of shepherding flocks of sheep as a full time occupation. At one time it was never thought that a female could manage to do this successfully, but they can, and do, make as good shepherds as the men.

When you travel out into the countryside in the Yorkshire and Durham Dales, or into the hill country, you are most likely to find sheep and probably suckler cattle, which are cows with their calves running with them at their feet. There are now over fifty breeds of sheep in this country, and all the different breeds have their own characteristics and peculiarities. Some sheep are pure bred, and others cross bred to get a particular lamb which some farmers prefer, and which may suit their farms. Some breeds have been developed in specific parts of the country, and in certain areas the name of the sheep is taken from the district where they were born and bred. There are various strains of sheep bred specifically for the hills, such as the Blackface, the Swaledales, the Cheviots, and the Herdwick, whereas in the south of England, Suffolk and South Down sheep are popular. Some very old breeds of sheep survive in Scotland and the Northern Isles today, notably the Soay sheep on St. Kilda, although there are no longer any pure bred Shetland sheep of the original strain. However, many of the characteristics of the old breed survived in the modern Shetland sheep. Later introductions, such as the Blackface, the Cheviots, and the Lewis Blackface sheep were developed as local breeds around the 1760's. The old Whiteface sheep, similar to the Shetland breed, were once common throughout Scotland, but declined in numbers with the introduction of the new strains, and had vanished by the 1860's. However, in Scotland, they still have numerous breeds of sheep which are all different.

Large scale sheep farming dates back to the Border abbeys in the twelfth century. Sheep were common elsewhere, and as well as the farmers' packs or flocks, they were often kept for family purposes. The wool provided clothing, and the milk was made into cheese, ewe milking being carried out in the Borders well into the last century. There are still quite a number of sheep flocks today which are still milked by hand as they were in the past. The cheese which is made by the sheep farmers is top quality, and is sold by some of the foremost stores throughout the country. The farmers' markets today are also good outlets where farmers can sell their products, such as sheep's or goat's cheese, ice cream, beef, lamb, milk, fruit juice, and breads etc..

Farmers' markets are not a new innovation. Farmers have sold their produce in the market place since the early 1900's, as well as at the feeing markets where the farm servants were hired or they were looking for a new farm. The farm workers were either hired for six or twelve months at a time, and many of them kept changing farms every six or twelve months, mainly because of the conditions on some of them, poor wages, long hours, and the poor food available on some farms. The food was every bit as important as the wages. The hirings were held every May and November in the town or village square. The talk among the farm workers at the hirings revolved around the type of meat house of this or that farm. When a labourer found a farm which was a good meat house he would stay there longer. Farm workers rarely left good meat houses.

Normally unmarried farm servants lived in the farmhouse and received their board and lodgings, plus a money wage. Married farm servants had a farm house or cottage, and their wages were partly in money and partly in kind. They were given a supply of milk every day, a weekly supply of oatmeal, and as many potatoes every six months as the family could eat, as well as as many turnips as they wanted.

Some of the single general farm workers and shepherds that lived in the farm bothy made their own food, and often ate oatmeal for breakfast, dinner and tea. It was easily prepared by pouring boiling water over the oatmeal, adding a pinch of salt, and also a drop of milk. A bothy would normally have two wooden box beds with straw mattresses and straw pillows. The furniture would have consisted of a fireplace, a chair, a box for clothes, a meal box, a washstand, a basin and slop pail. There was no electricity in many bothies, and the farm workers carried a paraffin lamp around with them at night. On a cold winter's night the bothy could be perishing, as there was nothing with which to light a fire. One farm worker once stated that his bothy was so cold that he went to bed with his "tacketty boots" on, and pulled the old grey blanket over his head to try to get to sleep. Most of these buildings were dull, depressing places, and were cold nearly all the year round. Some farm servants would have been much warmer lying in the stable among the horses, or with the cattle in the byre. Surprisingly, some of the bothies were still being used by shepherds and general farm workers right up until the 1970's.

One of the main difficulties of sheep husbandry has always been keeping the numbers of parasites under control. At one time sheep were smeared, the fleece was parted in parallel lines, and tar, or sometimes nicotine was mixed into a grease paste, and rubbed on the fleece. This made the fleece difficult to clean when being processed for spinning, and dipping was introduced. It is interesting to note that the early dips included mixtures of nicotine, soap or sulphur! All sheep now, by law, have to be completely dipped once a year in a chemical for so many seconds to kill any parasites which may cause Sheep Scab.

SWALEDALE SHEEP

The Swaledale sheep get their name from the valley of the river Swale, situated in the Yorkshire Dales. Sheep were carefully selected and bred on the Yorkshire moors for many years until there emerged the remarkable Swaledale sheep that we know today. They are an active and alert breed, and are a reasonable size to suit the moorland environment. Their hardiness is proved by their ability to live out of doors on the wild tough hills from where the breed originated.

As the Swaledale sheep travel on rough countryside, they have good strong feet and legs. They also have short broad teeth to eat the heather, moss and grass which is their main diet. Swaledale breeders have found from experience that the Swaledale ram will produce a hardy vigorous offspring which do very well on lowland farms where there is plenty of good grass.

The Swaledale ewe spends the first four to six years on the hills in the flock where she was born and bred, during which time she will produce flock replacements. The lambs will either be sold in the store market, or in the fat market ready for the butchers. After this period of her life on the hills, the ewe will then leave the hills and be bought by a lowland farmer who will cross her with either the Teeswater ram or the Bluefaced Leicester ram, which will produce the Masham and Mule gimmer lambs.

These females are supreme for lamb production, and do extremely well on the lowland farm.

BLACKFACE SHEEP

The Blackface sheep is one of the most common sheep in Britain today. This breed has spread all over the world today, wherever the climate and conditions are particularly hard. The Blackface breed of sheep is very hardy, and can stand up to, and survive, the worst of snowstorms on the open hills.

I can well remember the 1963 snowstorm in the north of England. The whole area was disrupted for months. There were night after night of very severe hard frosts, and the snow was so deep you couldn't travel around for weeks on the roads, until they were cleared. Out on the hills many sheep died of starvation, because some farmers ran out of fodder, and some of them couldn't even travel out onto the hills to feed the sheep because of the deep snow. The snow was two or three metres high on some roads, and this took some time to clear. Around the farm buildings and in some fields the snow drifted as high as six metres in some places.

One hill farmer that I went to see during this long, hard winter was in a desperate situation. He kept a flock of two hundred Blackface sheep, some of which were standing in deep snow which had frozen solid. It was a dreadful sight to see them. I helped the farmer to look for some of his sheep buried in the deep snow. Some of the Blackface sheep we found were still alive after being buried for several weeks! They somehow managed to get some air through the holes in the snow. Most of the sheep that we found alive were only skin and bone, and couldn't eat, and so they eventually died. We did, however, drag a few live sheep down through the snow, and they started to eat a little hay when we got them down into the farm buildings.

Apart from the Blackface sheep being such a hardy breed, they have a built-in instinct to remain on the hills where they have become settled. This instinct is known among the hill farmers as "Hefting". When the sheep become acclimatised to an area, they will not voluntarily leave the area, nor mix with sheep from another locality. This instinct reduces the need for fencing off the moors and hills. This is partly why you see so many open hills and moors, and why the sheep on the roads know exactly where they are, and in which direction they are going.

BLUEFACED LEICESTER SHEEP

The Bluefaced Leicester sheep was developed near Hexham in Northumberland at the beginning of the 1900's. They were bred to produce top quality cross-bred ewes from the native Blackface and Swaledale draft ewes.

At the beginning of the First World War the Bluefaced Leicester (or Hexham Leicester as it was then known), was a distinct breed in appearance. The breed became increasingly popular, and soon became established in the Dales in the 1920's and 1930's.

The Bluefaced Leicester ram is bred and used for the same purpose as it was developed, that is to produce top quality cross-bred ewes from draft mountain ewes. The cross is traditionally called the Mule.

Farmers from Northumberland visiting Mr. Joe Raine's Old Park Farm at Kirkoswald in Cumbria in 1988. The son, David Raine, is showing the farmers his Blueface Leicester tup, for which they paid over £5,000.

Farmers from Tynedale having a farm walk, looking at the sheep on Joe Raine's farm at Kirkoswald in Cumbria.

Two young Blueface Leicester tups on a farm in Cumberland.

A farm collie sent out to collect some sheep at Bishop Burton Agricultural College near Beverley , East Yorkshire.

Sheep being dipped with disinfectant to prevent sheep scab, at High Leam Farm, West Woodburn in Northumberland in 1981.

Sorting sheep out at Embley Farm in the north, getting the sheep ready for the local market.

Farmers from Heddon, near Newcastle-upon-Tyne, with their breed of pedigree Suffolk sheep.

Mr. Arthur Yeates from Northumberland, judging the Blueface Leicester sheep class at the 1980 Royal Highland Agricultural Show in Edinburgh.

Eddie Armstrong from Wooley High House Farm in Allendale in Northumberland, standing on top of a snow drift in the 1963 storm.

CHAPTER 6

SHEEP STELLS OR SHEEP FANKS

Sheep stells or sheep fanks of the Border country were introduced around 1760 by the Border shepherds, along with the drystone dykes. Sheep stells are dry stone circular walled buildings built on sheltered parts of the hills, where flocks of sheep can be gathered for protection during stormy weather. The stell walls are about 1.52 metres (5') high, and vary in diameter according to the flock size. Most stells in the north are circular, although occasionally some have been built in a square. In Ireland, some old stone stells were constructed in the shape of a cross, for it did not then matter in which direction the wind, rain or snow was blowing, as the sheep or cattle could go into one corner of the cruciform shape to find shelter and protection from the elements.

In the south of England, shepherds used to carry gait-hurdles around on their backs, these being made of woven willow and very light to carry, so that they could easily be made up into a pen. The stakes were also very light. Sheep could be gathered near lambing time, or could shelter from the wind and rain within the willow hurdles, which were sometimes covered with straw or canvas.

With open hillsides and common grazing in the mountains, hills and dales, the identification of sheep has always been a matter of importance. This, at one time, was done with lug marks, a pattern being clipped out of either one, or both the sheep's ears. Each district developed a complicated grammar of different shapes and combinations. This has been replaced in most areas with the use of a dye, or branding, although in some areas of Cumbria both lug marks and dye are used to mark a flock.

Before pigs became common in the late eighteenth century, a poor family, if lucky enough to eat meat, ate mutton.

A stone-built sheep stell on a sheltered part of a hill in the north.

Another type of sheep stell made with corrugated sheeting.

A Hogg hole.

The hole at the bottom of the stone wall, where sheep can pass from one pasture to another, is known as a Smout Hole, Hogg Hole, Sheep Run, Lunky Hole, or Sheep Smoose.
Note the sheep in the background .

CHAPTER 7

DRY STONE WALLING

When visitors travel north, or come into the Dales, they are often attracted to the patterns of the stone walls, how they are constructed, and how they snake up over the hill tops and down the valleys, and form different patterns.

The stone walls are very useful to landowners, farmers and shepherds, as they separate the land, hold livestock in their enclosures, and shelter sheep and wildlife which are out on the hills and moors. To the shepherd and his hill sheep, the stone walls are invaluable because at lambing time, in the spring of the year, as the rain sweeps across the hills, the sheep and lambs shelter by them. Many lambs' lives have been saved by them and the protection that they afford. When we get what is called a lambing storm in March or April, sometimes a strong wind blows with heavy rain, sleet or snow. The walls then come into their own as shelter belts.

During the winter, some hills in the north get much more snow than others, and it can be very rough out on the hillside for the sheep, cattle and wildlife. Some flocks of sheep know well in advance when a blizzard or stormy weather is coming, and they move down from the hills to the lower ground. The sheep do, sometimes, get caught out by the weather, but not very often.

Dry stone walls are built by craftsmen, and are as varied in their structure as in the types of stone with which they are built. There are three types of stone used in the north and in the Dales for building the walls; Whinstone, Limestone, and Sandstone. Most of those constructed in the Dales are formed from Limestone.

Constructing a stone wall is not as easy as it looks. I am busy repairing a stone wall on our farm which was hit by a car. About three metres of the wall has had to be rebuilt. First of all, the foundations of the wall are dug out a few inches, and then big flat stones are laid. These are known as the Footing stones or Foond. The wall is built up on two sides of the Footing stones, and the space between them is filled with small stones. These are known as the Hearting or Packing stones. About 1' 9" (fifty three centimetres) or so above ground, large flat stones are laid the full width of the wall, plus a little extra. These are known as Through Bands, which help to strengthen the structure, and tie it together. When the barricade is about 3' 2" (one metre) high, another band of Through Stones can be put through the wall, depending on how high the wall is going to be when completed. The two sides of the wall are built up to the Cover Stone, which is about 4' 6" (1.37 m.) from the ground. The Cover Band Stone sits flat on top of the wall to further bind it together, and to stop the rain getting into the structure. Finally, the Coping Stone, or Cap, forms the crown of the wall. They can be flat stones set on their end to stop sheep or cattle from nosing at it. However, most of the Coping Stones in the north are round shaped stones, which are packed tightly together. These round stones are very popular with some people, as they keep pinching them for their gardens!

Most field walls are built about 4' 6" (1.37 m.) to the Cover Stone, while many "march", or boundary walls or dykes are built a foot (30 cms.) or so higher. The dykes in Galloway in Scotland are constructed as single dykes, with one stone built on top of one another to make a stock proof barrier. A lot of the stones built on the Galloway dykes came from the river beds. It is not an easy job building one rounded stone on top of another, and there is definitely some expertise and craftsmanship in building these single dykes in Scotland to make them stock proof.

The reason why the walls were built "dry" was due to the fact that in the olden days, lime was the only thing to set the stones, and it was far too expensive to transport lime, sand and water away out onto the hill tops.

Why was there any need to build stone walls in the first place? Well, there have always been disputes over land, boundaries and trespass, and there always will be. The Act of Parliament of 1801 was known as the Enclosure Act. In 1792, an Act of Parliament gave farmers stints of pastures to enclose as the farmers were flourishing, and it was hoped that this would stop trespass with cattle. Stints are strips of land which were allocated to farmers next to common land. The stints could be sold to other stint holders, and the farmers could only graze so many sheep on a stint. The stone walls in the Dales and in the Lake District were all built after the Act of Parliament of 1801. Commissioners surveyed the lands to be enclosed by the walls, and the expense of building these was paid for by the tenant who was awarded the land to be enclosed.

The gathering of the stone to build the walls could be a big problem. In some areas there is a wealth of the material, which has been ploughed out of the fields over many years. In other places, such as out in the hills, the stone had to be quarried locally, and in fact was obtained by digging small outcrops of rocks as near as possible to where the wall was going to be built. These outcrop rocks can be seen on most hills where there are stone walls. The rock then had to be moved to where the wall was being constructed, and this was done with a horse and 'slipe' (sledge), which was no easy task. There is about one ton of stone in a yard of wall!

The men who were contracted to build the wall sometimes had the quarrying of the stone built into the price of constructing the wall. In the early 1900's, some stone wallers used to work in pairs, sometimes father and son would work together, and they could be paid anything from 2d. (1 p.) to 10 d. (4 p.) an hour. The wallers had to work long hours to make any money, working an average of fifteen hours a day. Wallers working on a farm or an estate would sometimes lodge with the shepherd or herdsmen, and were early in bed at night, and up early in the morning. They would sometimes, also, have to walk two or three miles out onto the hills before they reached the area where they were building the wall, and then, of course, had the same distance to walk back home at night after a long day's work! Another method of paying a waller was by "rood", which was seven yards (7.5 metres) of wall, and was a full day's work. If the waller, however, had to lead some of his stone a long distance, the price per "rood" could increase. The weather could be a real snag, as the wallers grudged leaving their work when it started to rain heavily. They would sit on their hunkers down beside the wall, and hope the rain would stop. It was not uncommon for them to carry an old piece of sailcloth or something like it, and rig it along the wall with a few stones for shelter when it became very wet and windy.

In some parts of the Dales where the land was too poor to grow hedges, stone walls were erected to separate the land, and keep sheep and cattle from straying. Many walls have certain interesting features, such as Hogg Holes, Smout Holes, and Lunky Holes, where sheep can pass from one pasture to another. There are also small holes at the base of the stone walls known as Rabbit Smoots, where rabbits and hares can pass from one pasture to another.

Some of these old walls are as dry as can be, and are often blown full of old leaves, or on the hills they can be full of grass. This can benefit the wildlife in the area, as they become snug places for birds and animals to shelter. Some of the walls I come across could be two hundred years old, and yet they have never been touched, proof of the craftsmanship that created them.

Filling the heart of the stone wall with small stones.

Building a stone wall at Haydon Bridge in Northumberland. The centre of the wall is filled with small stones, or rubble.

A stone wall with rounded coping stones along the top of the wall, which protects the structure beneath. The top stones are also known as topping stones, or the cap, or comb.

The small hole at the base of the wall is to let hares and rabbits pass from one pasture to another. It is known as a rabbit smoot, or a pen hole, or pop hole.

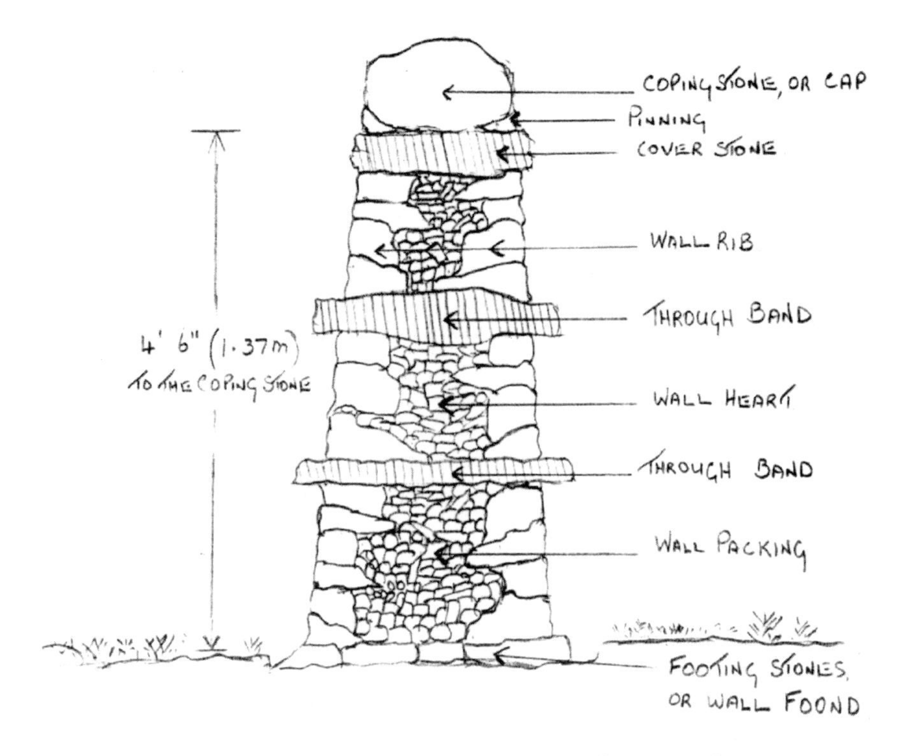

This illustration shows the construction and sections of a double dry stone wall in the north.
Field walls vary in size, and they are sometimes lower than boundary stone walls.

CHAPTER 8

THE CLYDESDALE HORSE

The Clydesdale horse is a lovely heavy draught horse which was developed in Scotland. It is thought that the breed started in the Lanarkshire area at about the middle of the eighteenth century, and as the River Clyde flows through Lanarkshire, this is how the horse got its name.

The story of the Clydesdale horse, however, begins a little before the middle of the eighteenth century. It appears that a Mr. John Paterson of Lochlyoch in Lanarkshire had a black Flemish stallion about the year 1715, and it is thought that this horse was originally brought there from England, but there is not much information about the stallion where it came from, and what it cost. John Paterson's Lochlyoch stud became quite famous. This Flemish stallion and the mares he bred from him were in great demand for many years to follow.

The 6th Duke of Hamilton, who acquired the title in 1742, imported a black Flemish stallion which was used on the mares of his farm tenants free of charge. So although the Clydesdale horse is reckoned to be a native breed of Scotland, the horse seems to have been bred originally from the black Flemish Dutch strain of horse.

The Clydesdale horse was bred and moulded from there on by the farmers of Lanarkshire. "Clydesdale" is the old name for this county. The characteristics and the type of horse were improved as a draught horse to meet the demands of the industrial development when the coal fields in the Lanarkshire area were being developed. The roads also needed to be improved, and much of the haulage of the materials was done by the native breed of horse. As the horse power increased it replaced the ox. As the breed developed, it soon gathered a more widespread reputation during the late eighteenth century, and the beginning of the nineteenth century. Many young Clydesdale colts and fillies were drafted from the fairs held in Lanarkshire, Rutherglen and elsewhere in the county into the Midland counties of England.

Because of the popularity of the breed, since about 1850 there became a steady export of Clydesdale horses to foreign countries such as Australia, Russia, New Zealand, United States of America, South America and Canada. The State of Victoria and the Province of Otago in particular got many very high class horses and mares.

The Clydesdale Horse Society was founded in June 1877, and the first volume of the Clydesdale Stud Book was published in December 1878. Exact data of the number of sales and exports before that were unavailable until the founding of the Stud Book, and after that the system of issuing export certificates by the Clydesdale Breed Society was established.

***

A SHORT STORY

In 1843, a William Eadie emigrated from Cambuslang in Scotland to Ontario. A few years later he returned to Scotland and he spotted a Clydesdale stallion called Campsie Jock. He struck a deal with the breeder and William took the stallion back to Canada. Campsie Jock was the first pure bred Clydesdale horse in Eastern Canada.

***

In 1912, when the *Titanic* was making its maiden voyage to New York in America, it hit an iceberg and was sunk. Along with a huge loss of life, there was also the loss of animals, for there were Clydesdale horses on board.

***

CLYDESDALES FOR TOWN AND COUNTRY

I would like to say that for town and country work the Clydesdale is a great animal, for his hardiness and handiness make him ideal for that kind of work. In the towns and cities, the Clydesdale geldings were used as draught horses, and were employed by the brewery trade, the bakery trade, the coal and rail industries, the council, and many other businesses as well as farming.

The late Bill Thompson from Prudhoe in Northumberland used to train the young Clydesdale geldings to haul the coal barges along the Manchester Canal. When the Second World War started in 1939, Bill didn't get called up because of the job he had hauling coals by barge to both Kellogg's and Mather & Plats etc. through Salford and into Manchester, which were classed as essential services supporting the war effort. Bill Thompson was a good business man and farmer. I used to meet him at the cattle market from time to time, and we had a bite of lunch together. He used to tell me that when a young Clydesdale gelding started working pulling the coal barges, if the young horse didn't take the strain on the barge and just charge forward, it would end up falling into the canal. Bill said when you had a young horse in the canal and it started to panic, it was sometimes a difficult job getting it out of the canal.

Clydesdale geldings worked on the streets of Glasgow for the brewery trade. A gelding, which was about five years old would cost between £100 and £140 in 1915. A five or six year old gelding was popular with the company for its ability to cope with street traffic, and the horses should weigh between seventeen and eighteen cwt (865-916 Kilos). The average load a horse would be expected to pull in the city would be between two and three tons on the flat, the cart weighing about twenty two cwt(1095 Kilos). When the brewery had to use Buchanan Street or West Nile Street which are on a fairly steep incline in Glasgow, they would use a trace horse, putting one horse in front of the other, both horses pulling the cart up the hill. Clydesdale geldings would last anything from ten to twelve years on street traffic, and they could travel between ten and twenty miles a day.

There have been a number of top quality Clydesdale horse breeders that have made a name for themselves over the years, such as James Kirkpatrick of Grange Mains Farm, Kilmarnock, and William Dunlop from Dunure Mains farm at Ayr. Both these men were very friendly and they had stallions travelling around the farms in the district. Some farmers would bring their mares to their farms to have them served by one of their top quality stallions, such as Dunure Footprint (15203).

This is a true story about how James Kirkpatrick and William Dunlop fell out over a Clydesdale stallion called Baron of Buchlyvie (11263), and they fought a battle in court over the stallion. Baron of Buchlyvie was born on the 16th May 1900, and was bred by William M'Keich, Woodend, Buchlyvie, Stirling. James Kirkpatrick bought Baron of Buchlyvie as a two year old Clydesdale stallion at Aberdeen Highland Show in 1902 for £700. James Kirkpatrick sold his friend William Dunlop a young stallion called Royal Carrick for £700, prior to his buying Baron of Buchlyvie. Unhappily, Royal Carrick died with William Dunlop at the beginning of the first season. James Kirkpatrick then offered William Dunlop a half share in Baron of Buchlyvie to compensate him for the loss of Royal Carrick, and William accepted the verbal offer.

About two weeks later, both men met in a pub in Ayr to square up over the half share in Baron of Buchlyvie. It was at this meeting that a misunderstanding arose between the two men. William thought £1000 was for his half share of the stallion, while James wanted £2000 for William's half of the stallion. The deal was called off.

In the spring of 1904 James Kirkpatrick sent £300 to William Dunlop, which was half the stud fees for the time Baron of Buchlyvie spent at Craigie Mains farm in 1903. Why James sent William these fees when the misunderstanding still hadn't been resolved, is a mystery. Baron of Buchlyvie was then delivered to Dunure Mains early in 1904, and the following year William Dunlop sent £250 to James as half the stud fees, and this was the last payment made between them. The stallion was left at Dunure Mains until 1908, when James then asked William for half the stud fees for the last three seasons. William replied that he didn't owe him any stud fees whatsoever. James was somewhat annoyed with William's remarks, and he took him to the Court of Sessions in Scotland, where Lord Sherrington, at the hearing, said that James still owned a half share in Baron of Buchlyvie. William Dunlop then successfully appealed to the Inner House of the Court of Sessions. James Kirkpatrick then decided to take the case to the House of Lords, which overturned the appeal decision, and confirmed Lord Sherrington's original verdict. Both men were now in an awkward dilemma, as they both each owned a half share in the stallion. It was then decided through both men's agents to put the stallion up for sale.

On December 14th 1911, a special sale was arranged at Ayr cattle market to sell the stallion. Between four and five thousand farmers and Clydesdale breeders attended the sale from all over the country. William Dunlop bought the stallion for a world record price of £9,500.

Three years later, Baron of Buchlyvie was kicked by a mare, and his front leg was broken. He was then destroyed, and William Dunlop buried the stallion in his rose garden at Dunure Mains Farm. Four years later, the remains of the stallion were dug up, and were sent to the Glasgow Museum at Kelvingrove Park, where the skeleton of the stallion was put on display. It is still in the museum today. Several years after the sale of Baron of Buchlyvie, Mr. Butler, a mutual friend of both men, brought both James Kirkpatrick and William Dunlop together again, and they remained friends until they died a few years later.

***

Before, and well after the Second World War, the Clydesdale horse in the northern half of the country, did all the heavy work. At the same time, the tractor, also known as the Iron Horse, was also being developed. The history of the tractor in Britain dates back to about 1896. The first company to achieve commercial success with it was Richard Hornsby and Sons Limited, of Grantham, Lincolnshire. The first tractor that I can remember coming onto our farm was a green Fordson, with iron lug wheels, during the early 1950's.

As the tractors kept developing, some of the big farms quickly changed to the mechanical power, and the heavy draught horses started to decline on the farms. There were, and still are, farmers in this country that prefer to use the Clydesdale breed, or other heavy draught horse breeds, to work their farms, and have never changed. It is thought that a pair of Clydesdale horses could plough about one acre of land a day from about eight o'clock in the morning until about four o'clock in the afternoon on fairly good soil. There are also a lot of farmers that keep a Clydesdale horse or two just as a hobby, and they take them to the various agricultural shows mainly around the local community.

The diseases with heavy draught horses can be due to bad management, or it can be hereditary within the breed. Clydesdale horses working on farms or hauling timber in the woods sometimes develop what is called "Mud Fever", which is quite a similar thing to a human having chapped hands. The horses get inflammation of the legs and heels, and this is caused by the horse working in wet muddy conditions for long periods. "Grease" is caused by dirty stables. The horse gets cracked heels, and these become filled with a thick foul smelling cheesy type of material which very often discharges. "Side-Bone" is another disease which can make the horse lame, and is a bony formation of either lateral cartilage of the horse's foot. This disease can keep the horse off work for a few days, and is best treated by the vet. "Monday Morning Complaint" is when the horse's legs become inflamed, mainly the hind legs, and is caused by the horse standing idle during the weekends, or by over feeding. The horse may become lame.

A pair of Clydesdale horses ploughing a stubble field. The man and the horses walk about eleven miles a day.

Two pairs of Clydesdale horses ploughing a stubble field. One man and a pair of horses can plough about one acre (0.40 hectares) of land a day.

A pair of Clydesdale horses returning from the fields at the end of the day.

In 1987, an open ploughing match was held at Morpeth in Northumberland. Here a pair of heavy horses are being led, while the ploughman is trying to plough a straight furrow.

Working the land can be heavy work. The tea break is always welcome.

A pair of Clydesdale horses ploughing a grass field were often followed by the Black Headed gulls.

This instrument is known as a gagging bit. It is used for keeping a horse's mouth open while its teeth are being filed, or it is being given some medicine.

This is a horse and farrier made with flowers, a lovely piece of artistic work. It was on display at the 1987 Royal Highland Show at Edinburgh.

Billy Thompson on a Cumberland farm, cutting corn with a binder and a pair of Clydesdale horses.

Cutting grass for hay with a pair of heavy horses on a farm near Harrogate. One man with a pair of horses could cut about six acres (two hectares) of grass a day.

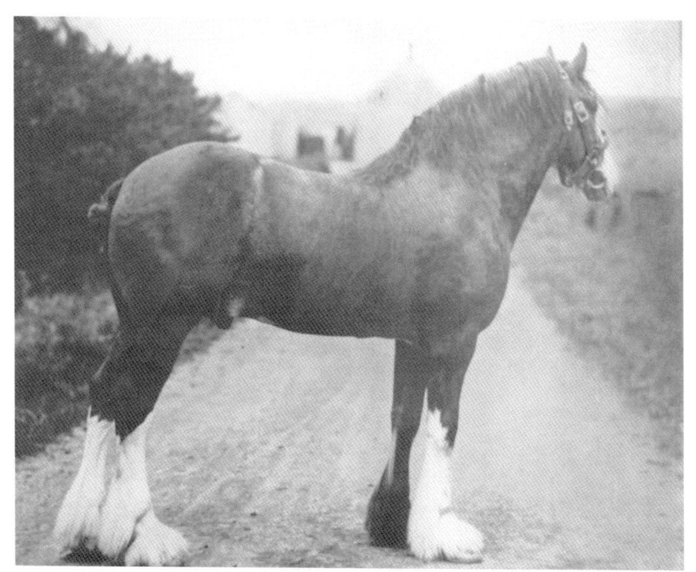

The famous Clydesdale stallion, Baron of Buchlyvie, sold for £9,500 on 14th December 1911 at Ayr Cattle Market.

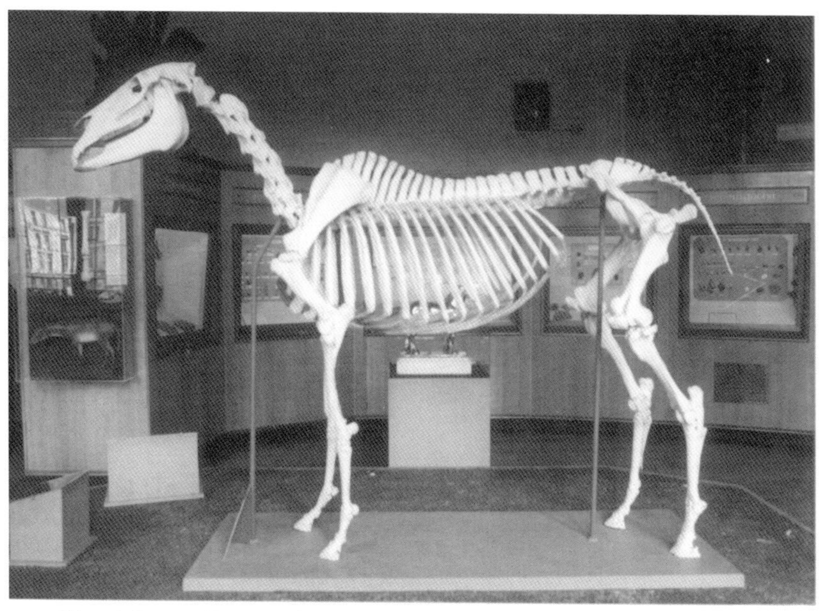

The skeleton of Baron of Buchlyvie, which is in the Glasgow Museum at Kelvingrove Park in Glasgow.

CHAPTER 9

THE RED DEER

If someone asked you what Scotland was most famous for, most people would, I think, say hills, mountains, lakes, salmon, and of course, the Red Deer.

THE HIGHLAND RED DEER

I will tell you first of all a little bit about the Scottish Red Deer, but it is about farming the smaller Red Deer in Scotland that I will give you the most information.

The Scottish Red Deer roam the hills in herds, females and males staying in separate packs. They are by nature a forest animal. In the Highlands they live in what are known as "Deer Forests", which are moors and mountains. In the winter months the deer come down from the high hills. They normally feed in the early morning or evening, and they eat grass, young heather shoots, moss, shoots of trees, leaves, and seaweed. You may be able to see the deer in the late afternoons in the winter months. They pay little heed to moving cars, and will just carry on feeding unless a car door is slammed, or a car stops suddenly.

The young deer are known as calves, and are born at about the end of May or early June. Their hair is dappled with white spots which disappear when they grow their first winter coat in two or three months time. During the first week, the young calf is unable to stand and the hind (female deer) hides her calf in bracken or heather. She leaves it and returns to suckle it later. The calf stays with its mother for about two years, and when it is about three years old it is ready to have a family of its own.

The horns of the young males first appear during the second year, and then only as frontal knobs known as knobblers. The next year a pair of single spiked horns appear, and the following year another point is added, and this continues until the full number is reached. Some stags never develop horns, and are known as "Hummels". A twelve pointer deer is a "Royal", and a fourteen pointer is known as an "Imperial". The stags lose their antlers in March and April, and often eat the cast off horn. The new antlers start to grow immediately, and are covered with a sensitive velvet. The deer then start to get pestered with flies, and they soon move back out onto the mountain tops. The weight and size of a full grown stag, which stands about 4' (123 cms.) to the shoulder, varies a great deal according to its feeding. Park Red Deer, which are fed throughout the winter may weigh as much as 30 stone (200 kg.), while a good Highland stag may weigh about 18 stone (150 kg.) or less, depending on the time of year.

The mating season, or the "rut" occurs during the autumn, when the stags can be heard roaring challenges to one another. The stags round up as many hinds as they can. Fights do break out among the stags, but they normally just end up with a few scratches and bumps.

The deer are stalked and shot with a rifle, and are not hunted with packs of hounds. The shooting season for stags is from 1st July to the 30th October, and for hinds from the 31st October until the 15th February. Stalking deer and shooting them requires a lot of skill and patience. My father used to tell me that the Red Deer could hear or smell you over a mile away, and you always had to keep the wind on your face.

FARMING THE SMALLER RED DEER

Up in the Galloway area of Scotland, George Osbourne farms the small Red Deer at Barhourie Deer Farm, near Kipford. The farm is 120 acres, and belongs to George, who is Polish. He came to England after the Second World War. He has always been interested in deer, and has had various jobs in England. His last employment was managing a deer farm in the south of England.

In 1981, Barhourie Farm came up for sale. George and his wife bought it for £40,000, which seemed very cheap, but the land in that area can be fairly rocky. Because the farm is in the less favoured situation, it gets subsidised. There were no farm buildings, water, or electricity, and George had to start from scratch, putting up some farm buildings, and getting water and electricity installed at the buildings. It was very hard work. When the buildings were completed, he started to buy in some Red Deer breeding stock from Charlcote Park at Stratford-on-Avon. (Incidentally, this is where William Shakespeare was arrested for poaching!). These Red Deer are smaller than the Scottish Highland Red Deer.

George now has a herd of two hundred and fifty deer, and he reckons to have one stag to about forty hinds. This, George said, was just an average of stags to hinds. At the "rut" he said that some stags will end up with more hinds than others.

The deer are butchered on the farm, and only the stags are sold as venison when they are between a year and eighteen months old, and kill out at about 120 lbs. (55 kg.). George selects a stag for venison and shoots it in the field with a rifle. He then takes the deer into a small butchering house he has had built, skins and cuts it up, and puts it in the freezer. The venison cuts vary quite a bit in price, from between £4 and £5 a lb. (450 g.). His butchering house was spotless.

The young yearly hinds are sold as breeding stock for £400 each, and a good breeding stag can cost around £2,500. The hinds are given a feed of lanzern nuts in a trough during the summer when they have a calf at foot. They are also fed potatoes during the winter, with a bit of good hay. The deer are very unpredictable during the rutting season, which starts in October and can last for two months and George said one had to be very careful.

One thing I learned about deer farming is that you can't drive the deer as you can cattle or sheep, as the deer just split. The day before I arrived at the farm, a young stag had jumped out of the enclosure. The next morning, when I was at the farm, George and I were standing talking in the farm yard, when this young stag came trotting back up the farm road. George quickly opened the enclosure gate, and we both guided the stag back into the pasture.

George said he can sell all the venison he kills. People order it by 'phone, or they call at the farm for it. I had a really good day there, and the venison was quite tasty as well. Venison, like other wild game such as pheasant and grouse, needs tarting up with sauces and gravy, and is delicious with wine.

Mr. George Osbourne from Barhourie Deer Farm on the Galloway coast in the south west of Scotland. George has a herd of about two hundred and fifty deer. He reckons on about one stag to forty hinds.

George Osbourne from Barhourie Deer Farm near Kipford on the Galloway coast in Scotland. George is feeding the deer with a few apples, which they really enjoy!

CHAPTER 10

HOW FARMING HAS CHANGED SINCE THE 1960'S

There are about 250,000 farms or holdings in Britain which vary in size a great deal, from about a one acre holding (0.40 hectare) to farms with a few thousand acres or hectares. The purpose of the agricultural industry is to produce food to feed the population in this country, and to export anything surplus to requirements, when there is any. The farmers in Britain, however, don't produce sufficient food to feed the populace, and between twenty five and thirty per cent of the food is imported every year because of various trade agreements. About fifty per cent of the farmers in Britain own their farms, and the other half are tenant farmers. Many farming families have husbanded the same land for generations, either as tenants or owner occupiers.

There is a vast range of different types of farms in Britain. These can be beef producing, stock rearing, pig, dairy, sheep, poultry, or fruit and vegetable farms, as well as those that grow a mixture of cereals. Due to the fact that the climate in the north of the country is much colder and damper than it is in the south, farming in the north can be a fortnight or three weeks behind the south with their products, and stock and crops are chosen for the different areas to cope with the nuances of our climate.

In some of the Yorkshire and Durham dales there are many small family farms, and life on the these has not been easy, although somehow many of them have survived. I started working among them at the end of the 1950's, and witnessed their struggles, and how hard life was for many of them. It was all manual work on the farms at the time, and the farmers were just recovering from the Second World War. There was little mechanisation, no computers, and very few combine harvesters. There were, however, plenty of farm workers, and they had to work very long hours at the busy times.

It wasn't really until the 1960's that the expansion in agriculture took off. At the time, there were Ministry grants available to amalgamate the land, and many small farms were combined to make much bigger farms or units. Because of this, thousands of farmsteads disappeared during the 1960's and 1970's, and this had far-reaching effects, as these small farms were a stepping stone for young people who wished to get started in farming on their own. What have we today? It is nearly impossible for young people to begin a career in farming. Even if they are given the farm rent free, they would still need a lot of money for the cost of agricultural machinery, stock, a milking parlour, and other working equipment.

As farming in the 1960's was expanding, there were a lot of farm changes at the May and November terms. The May term was the busiest time for the dairy farmers. I sometimes had as many as thirty farm changes at this term to register and license.

As the farms were getting bigger at this time, so was the agricultural machinery which was necessary to manage such large areas. Many farm field gates, and the entrances to many farms were just not big enough to cope with the new modern agricultural machinery. Most of the farm buildings were not suitable, either, for the modern tractors. As the dairy herd numbers started to increase in the early 1970's on many farms, the cow sheds weren't suitable for the development in the herd size. At this time the average dairy herd size was thirty cows, but by the early 1990's the typical herd size had increased to one hundred cows. Today this figure has doubled to about two hundred cows.

To cope with the bigger agricultural machinery and the expanding dairy herds of the 1960's and 1970's the farmers started pulling out hedge rows, demolishing stone walls, and grubbing out trees and plantations to grow and produce more food, which they were encouraged to do by the Government. The dairy farmers began converting their farm buildings, or built new ones so that they could install milking parlours, and also to accommodate the increasing cow numbers. The modern, or altered buildings provided a more efficient method of handling large numbers of dairy cows. As these herds grew bigger, the farmers started to keep records of the cows, and some farm offices were built or converted in the farm buildings. The computer age was arriving, and some farmers were quick to make use of it.

With the vast changes and modernisation of the farming industry in the 1960's and 1970's, thousands of farm workers discovered that they were no longer needed on the farms, and started to leave the land in droves. This is still continuing today as the industry keeps changing. Those that remain on the farms need to be highly skilled to manage to work with the ever developing modern agricultural machinery.

The farmers are a group of very independent characters, and mostly struggle along doing their own thing. However, as farming has become more specialised and mechanised, the cost of some of the new agricultural machinery is beyond many farmers' pockets, and in the last ten years or so, there have been groups of farmers forming machinery syndicates. There is one group of farmers in the north who are based in County Durham, and they have a machinery syndicate running. However, many farmers buy second hand agricultural machinery where they can at farm sales, and from merchants. There is also a syndicate of grain farmers in the north, based on the east coast in Northumberland, and they buy the grain from farmer members, and sell it and ship it abroad to Holland, Germany, Denmark or France, wherever the group can sell the grain. Some of the grain farmers use most of their crops on the farm, and sell what surplus they may to a grain merchant. It is mainly the bigger farmers that form the syndicate groups, and there are not many of them. Small farmers don't produce enough of any one commodity to be in a marketing syndicate.

At this time of year (late spring, early summer), the farmers make their first cut of grass which is made into silage. Some neighbouring farmers help one another at this time. One farmer will cut the grass, while maybe another two farmers will pick it up with a tractor and trailer, and lead it into a silage pit or silage tower. Silage making is all done by machinery. The silage that is put into a silage pit is rolled with a tractor to get all the air out of the pit, and this gives a better fermentation. When the pit is full of grass and rolled for the last time it is then covered with plastic sheeting, and old tyres are placed on top of the sheet to help the grass to ferment more quickly.

This is how the farmer next to us made silage. He cut the grass and left it to dry for two days or so. Then he carted the grass into the hay shed and put a layer of the grass in one section of the shed. He then covered this with treacle. Then he put another layer of grass on top of the first layer, plus another covering of treacle, and so on. I have seen at least thirty layers of grass and treacle in the hay shed! You could buy the treacle in forty gallon barrels. When the grass fermented with it, it made top quality silage. The dairy cows went for this silage, and there was never any left!

Silage has been made in pits and clamps for as long as I can remember. Then, in the 1980's, the silage grass was rolled into a big bale by machine, and then put into a black plastic bag, and tied at the end to make it air tight and help the fermentation. Grass is best cut for silage when it is dry and the sun is shining, because it is the sun that produces the sugar in the cut grass. I can remember a few years ago I cut a field of grass for hay. The weather wasn't very good, and I thought that when I got the grass as dry as I could, I had it baled and put into silage bags. I thought I might have some decent silage anyway. When I came to use it, I opened the first bag, and could hardly believe my eyes. The silage was mouldy and rotten in places. The reason for this was because I didn't tie the bags properly, and the air got into the bag and stopped the fermentation, causing the silage just to rot. I ended up trying to burn all thirty bags of it, and that took several weeks. Making good silage is a skill, and should be made when the weather is right.

Many farmers' wives have a job away from the farm, for they have to make ends meet. Long gone are the days when the farmer's wife used to keep a few chickens or ducks, and she used to feed all the calves. A lot of farmers' wives also used to help in the hay field, but again this is all changed, as most of the hay making is now done by tractor and machines. As the agricultural industry was developing and changing in the 1970's and 1980's and things were getting mechanised, the farmers' wives were getting made redundant on the farm. The price of mechanisation was costly, and some farmers had to increase their overdraft at the bank. Then when farm prices were not very good, some farmers couldn't make enough to cover their overheads, and their wives had to look for a job to live. Many farmers' wives have gone back to work. I know a few that have returned to nursing, banking, and working at the cattle markets. The cattle markets operate one or two days a week. One day they may sell store cattle and sheep, and another day they may sell fat cattle and sheep. Most towns have a cattle market every week selling something or another. Some farmers and their wives work in these markets part time, which helps them to make ends meet.

Where does the agriculture industry go from here? I was against the amalgamation of the thousands of small farms in the middle to late twentieth century. I used to tell my colleagues in the Ministry of Agriculture at the time that these small farms were there to encourage young people to start farming on their own. Some of the Ministry officials agreed with me, whilst other thought that I was old-fashioned. I wonder? With so many small farms disappearing, and larger ones evolving, the appearance of the countryside has changed. Now, thousands of farm buildings, such as barns, cart sheds, and cow byres have been converted into housing. There are now vast acres or hectares of agricultural land without any farm buildings. I think that we may have to start heading back to where we were, before the "improvements" of the 1960's. With such wide areas of land with no farm buildings, in years to come those young people who wish to start farming may have to buy or rent a few acres or hectares of land, construct some farm buildings, and so the process will start all over again!

Jim Herdman, who farmed at Middle Shield Farm near Hexham for many years.

Two seventeen gallon milk cans, or milk churns, which were used on some farms in the north up to the 1950's. The farmers used to take their milk to the station in milk churns like this.

Corn being threshed at a farm in the north.

An Oliver tractor driving a threshing machine. There were normally about a dozen people working at the farm on a threshing day.

A butter churn on display at the agricultural museum just outside York. The butter churn was once used on many dairy farms.

A single furrow horse plough.

An old reaper, or grass cutting bar. One man with a pair of horses could cut about six acres (two hectares) of grass a day with this machine.

Some ladies working in a hay field on a farm in Yorkshire in the 1950's.

An old stone farm house and farm buildings. The door above the cowshed is where hay is stored for the cattle which are housed for the winter. The roof has stone slates.

Bob Stoddart ploughing with his grey Fergie tractor, which was made in 1946, at a tractor ploughing match.

At the same ploughing match at Morpeth in 1987. An old Fordson tractor in action.

Lewis Smith, centre, a farm worker from Fenwick Towers Farm, Stamfordham, being presented with the N.F.U. Long Service Award by Herbie Snelgar in December 1983. Dick Stappard, his employer, is on the left.

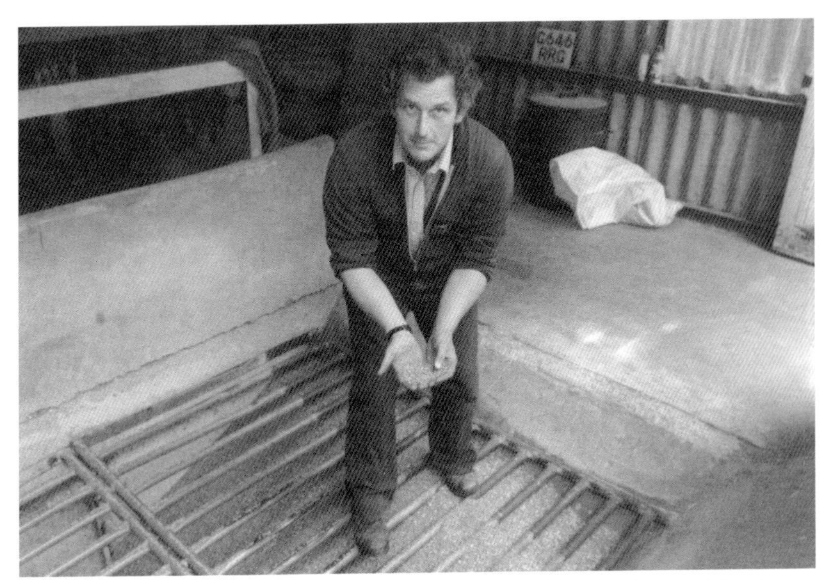

Mr. Glen Sanderson from Eshott South Farm in mid Northumberland, having a look at his barley crop as it goes into the grain dryer.

Small hay kiles or quiles on a farm in the Dales. The kiles help to dry the hay.

A tractor making the first cut of grass for silage.

Grass which can be wrapped or bagged for winter fodder as silage.

SHOWS

Young Cameron and Alexander Baty from North Acomb Farm at Stocksfield, with their calves at the Holstein Friesian Breeders Calf Show in Hexham Cattle Market.

Dairy cows in the show ring at the Northumberland Agricultural Show at Corbridge in 1985.

Young farmers judging a dairy cow for conformation on a dairy farm at Heddon on the Wall at Newcastle.

An Edinburgh farmer showing his pony at the Highland Show.

A small farm milk pasteurising plant on display at the 1981 Royal Highland Show in Edinburgh.

A programmable out of parlour feeding system for cattle. This is the Hunday Prize winning entry in the Barclays Bank new equipment awards at the 1979 dairy farming event. This shows the development of milking and feeding equipment.

Robin Baty from North Acomb Farm near Stocksfield, with his farm shop at Tynedale Show. Robin and his wife sell a wide range of quality dairy and country products in their farm shop.

A tractor demonstration at the Royal Highland Show in Edinburgh. The tractor is lifting a large bale of hay, demonstrating how high it could stack the hay.

CHAPTER 11

FOOT AND MOUTH DISEASE

There are some farm animal diseases which crop up every year among the cattle, pigs and sheep that the farmers and vets can normally sort out for the farm animals, such as mastitis, foot problems and sheep scab, etc.. When a disease like Foot and Mouth crops up on a farm, then it is a very different story, and needs drastic action to get the disease under control. The action we take in this country is to isolate the animal as quickly as possible, and then have the animal slaughtered and burned or buried in lime.

Foot and Mouth is an awful disease to strike the farming community. Any cloven hoofed animal, such as cattle, sheep or pigs are prone to this disease. It is an infectious viral condition which causes a fever in the animal, followed by blisters in the mouth and on the feet. It is more contagious than any other disease affecting animals, and it spreads like wild fire if uncontrolled. In the past the disease had been prevalent on the continent of Europe. It is thought that the infection was brought to this country by airborne carriage of the virus under favourable weather conditions, or through imported meat already infected with the virus. The early stages of the disease can be very difficult to detect, and it needs the vet to inspect the animal, and for the farmer to receive his or her advice on a suspected case.

Signs of the Foot and Mouth disease to be looked for in cattle can be shivering, slobbering, tender and sore feet, sores and blisters on the feet, and reduced milk yields. Symptoms of the disease in sheep include lameness, the sheep seeming "off colour", not moving when made to stand, may often lie down, and blisters on the hoof and mouth. Indications to look for in pigs can include sudden lameness, not eating, blisters on the snout or tongue, blisters on the upper edge of the hoof where the skin and horn meet, and the infected pigs also have a tendency to lie down. The Foot and Mouth virus can be present in great amounts in the fluid from the blisters, and it can also be present in the saliva, exhaled in the air, and extant in the milk and dung. Any of these can be a source of infection to other stock. Heat, sunlight, and disinfectants will kill the virus, whereas the cold and darkness will tend to keep it alive, and it can survive for long periods in these conditions.

Animals can pick up the Foot and Mouth virus either by direct or indirect contact with the infected animal or by contact with foodstuffs which have been contaminated by the infected animal. Cattle trucks, lorries, market places, and loading ramps with which tainted creatures have been in contact, are sources of infection until they have been thoroughly washed and disinfected. Roads may also become contaminated with the virus, which may be picked up and carried on the wheels of delivery lorries and milk tankers, etc.. Any person who has been in touch with an infected animal can carry the infected material until they have been cleaned down with disinfectants.

The owner of any suspected animal or carcass of the Foot and Mouth disease must report it immediately to their local agricultural ministry office or the police. If a case is confirmed on a farm, then an "Infected Area" will be imposed which extends to a minimum of six miles (ten kilometres) around the infected farm. The size of the area may need to be increased if the prevailing weather conditions are likely to spread the disease further. Movements on or off the contaminated farm will be controlled by the police or the local authority inspector. The Ministry of Agriculture will thoroughly disinfect all the farm premises. Other suspected animals on the farm may also be valued and slaughtered immediately, as well as the stricken animal or animals.

Protecting farms and live stock against the Foot and Mouth virus is very difficult. It is necessary to step up the hygiene standards and movements off and on farms with approved disinfectants. All vehicles entering and leaving should have their wheels sprayed with disinfectants. Farmers should keep their live stock away from household waste, swill and bones. Stopping all non-essential vehicles and visitors from coming onto the farm is essential, and farmers should arrange where ever possible to pick up any supplies and post at their farm gate. When there is a scare that Foot and Mouth Disease may be in the region, any extra hygiene standards people take on their farms will be a great help to try and stop any spread of this horrible virus.

Although the cattle in some South American countries, such as Argentina and Brazil, are vaccinated against the disease, the cattle in this country remain unvaccinated. There are arguments for and against this, but I believe that when things are working alright, one shouldn't tamper with them. However, on the other hand, when the hygiene of the animals is partly neglected, and there are too many crammed together, there is a high chance that this will cause a problem. I shall say more about this when I tell you about the 2001 and 2002 Foot and Mouth outbreak in this county.

At the time of the 1967 and 1968 Foot and Mouth outbreak I was working for the Milk Marketing Board as a field officer covering the north of England and along the Scottish Border. It was a dreadful time. Our regional office was then situated at Gosforth, and the Ministry of Agricultural Regional Offices were at Kenton Bar at Newcastle, and Northallerton in North Yorkshire. Most of the farms affected by the disease in the north were situated in Northumberland.

The Ministry of Agricultural staff and the Milk Marketing Board staff all clicked into gear together to try to beat this horrible disease. They had been rehearsing for emergencies like Foot and Mouth for some time. All the farm visits were stopped. The milk from the restricted farms was either fed to the live stock on the farm, poured down the drain, or put into a slurry pit. The milk tankers used disinfectants at each farm, and the milk churns were all brought to the roadside for the churn wagons to collect. These wagons also carried a supply of disinfectants.

I visited the Councils up and down the east coast to try to agree about disposal points where milk could be put into the sewerage from the tankers, and put into the sea. I had also arranged various collection points for milk churns, so that vehicles could be kept off the farms where possible, in an effort to try and minimise the spread of the Foot and Mouth disease.

In 1970, after this outbreak of Foot and Mouth in the north, a proper MILK CODE OF PRACTICE for the infection was set up by the Ministry of Agriculture to cover the dairy industry.

The next outbreak appeared in Brittany in France in 1981, and then another flare-up appeared just after this on the Isle of Wight. Movement restrictions in large parts of the south for farmers were introduced. It was thought that the Foot and Mouth virus could have been carried to this country by the wind from Brittany. The milk was not allowed to leave the restricted area for some time, whereas beforehand it had been sent to London every day. There were approximately five hundred and eighty five cattle slaughtered on the Isle of Wight in 1981.

****

The outbreak of Foot and Mouth in this country during the years 2001 and 2002 was just indescribable. The heartache the disease caused to so many farmers and business families alike was very difficult to watch and understand.

So why did this horrible disease get in among our livestock again, and cause such havoc? The finger was pointed at a pig farmer at Heddon on the Wall near Newcastle for his neglect and lack of hygiene among his pigs, and for not boiling his swill properly which he collected from the school, hospital, and bakery in the area in order to feed his pigs. I was involved with the owner and pig farmer a lot before the accused farmer got the tenancy of his farm. I used to sell him one or two tanker loads of skim milk for his pigs from time to time. The farm and pigs were always very clean, and his hygiene standards were very good. It is very easy to criticise people and organisations when a disease breaks out so widely in this country.

My thoughts and experience of this latest outbreak of Foot and Mouth in 2001 and 2002, and my observations in the past, have made me draw to the conclusion that the Government, which are not agriculturally minded, should have stepped up a gear at the beginning of the outbreak, and clamped down on all movements of livestock throughout the country. At the same time, there were far too many animals in the country which were being moved around without any thoughts about their hygiene. Animals and birds kept too intensively together, and with poor standards of cleanliness, definitely contribute to the spread of this disease. Some farmers are so greedy for land that all they want to do is fill the fields with animals that nobody wants, with little thoughts to their sanitary conditions.

Let us hope that the agricultural industry has learned something from this latest appalling outbreak. I hope that the movement of farm animals will be restricted, and that the build up of animals in the country doesn't continue to develop to the extent that it has done over the past few years.

CHAPTER 12

FARM SALES

There comes a time in our lives when everyone has to give up work, or sell their business, for various reasons. It may be because we have reached retiring age, or due to ill health. Also, sometimes the sons don't wish to follow on with the family business. In recent years, there have been other contributory factors, as I shall write about later in the chapter.

Farm sales in the north of England are usually held in May, and some cattle markets hold farm auctions nearly every year. Most of them are held in the open air, and the auctioneers have to proceed in all weather conditions. I attended a local farm sale in May last year, and the weather ranged from hail storms, to heavy rain, strong winds, and lovely sunshine. I felt a bit sorry for the auctioneers at times, but they carried on regardless, and did a good job for the farmers.

With over two hundred thousand farms in the country, with approximately half of them rented, most of the sales are for tenanted farms, although a few are for owner occupiers. Sometimes they only sell their live stock and farm implements, and continue to stay on in the farm house, letting off the land as grass parks. However, this usually only happens when the farmer owns the land and house, and very few tenants carry on living in their rented accommodation.

Farming has been declining over the past ten years or so, as the population has been changing its eating habits. Rice and pasta type dishes and fast foods seem to be more popular than roast beef and Yorkshire puddings. The trouble began when there was too much food being produced, and when the food is not being sold and eaten, then the amount of food production has to be reduced. When the farmers have to cut back on their products, some can't make ends meet, so they look for alternatives to make a living. The farm rents and farm overheads still have to be met. The recent outbreak of Foot and Mouth disease reduced our beef, sheep and pig population by millions, due to the animals that had to be slaughtered and burnt, yet I have not heard anybody say that there is a shortage of food. There have been thousands of dairy herds also which have been reduced since the outbreak of this terrible disease. Many of them will not come back into the dairy industry, and they will have to find something else to do with their farms. The milk quota which was introduced to the dairy farmers in 1984 will end on 31st March 2004.

Some farmers do adapt and try other methods of making a living from their farms, such as fruit picking, selling fruit and vegetables, bed and breakfast, horse livery, eggs and farm shops, to name but a few. A number of farms have also opened their gates to the general public. Some farms are only open in the summer, and other farms open all the year round, depending on the weather. Many farmers have converted their farm buildings into holiday lets. A few years ago, one in six farms had opened their barns and bedrooms for bed and breakfast customers, and this number keeps increasing. Farm holiday lets seem to have also increased over the years, with farm cottage holidays becoming very popular all over the country.

Some farmers can adapt to change more easily than others because of where their farms are situated. Some farms which are near to villages, towns and cities can change more easily than those away out on the hills. Many hill farms are mostly suitable for sheep and suckler cows, and the farmers can't just change to some other method of farming because of the area, and the terrain in which they live. As the agricultural industry keeps changing, it is very important to keep these farmers and their families out on the hills to produce good food for us all, and to look after the countryside, which they are so good at doing. There are some farms on the edge of the moors in the north that still produce a little drop of milk, but they are getting less and less as the cost of collecting a small quantity of milk can be very expensive.

The one thing that has been thriving over the years in the countryside is farm theft. The chain saws, tools, quad bikes, and anything that is worth any money is a target for the thief. I lock up all my farm doors and sheds just as a precaution after I lost some stack rope. There is big money in farm machinery, parts and spares, and the thieves know this. At one time the fertilisers and other animal feed stuffs were kept outside in the open. Now, if you leave anything mobile around like these commodities, you will be sure to lose them. Horse harness is another thing that the thieves will soon steal if they get the chance. I have known friends having trailers and horse boxes stolen. One particular dairy farming friend of mine had a tool shed broken into during the night, and £3000 worth of tools were stolen. The locks on the shed doors were forced open with pinch bars.

Sheep and cattle rustling is another problem that some farmers have to put up with. I am sure there are sheep and cattle rustled from the farms that are never reported.

Due to all the above issues, farm sales have become more frequent. We seem to be a nation for hoarding junk, and I am no different from anyone else when it comes to keeping things. The junk that is put out to be sold at some farm sales is incredible, yet there are always people there to buy it. If you visited a farmer with some of the rubbish and offered to sell it to him, he would definitely tell you to go away! However, at farm sales, anything seems to sell.

At one farm auction that I attended, there were so many boxes of nails and staples for sale that the whole country could have been fenced in, and there would still have been some nails left over! I think that this farmer must have bought some of them at sales in the past, as they were all in good condition. At the farm sales, there are always a lot of good things which are always in demand, but although bargains are there to be found, you have to know roughly what the new price of the article is beforehand. Some people buy something because they think it is cheap, which it isn't if they don't need it. A farm sale can be a traumatic time for a farmer and his family, especially if they have farmed there for a very long time. I knew one farmer who was very distressed at having to leave his home, as he had lived there for over forty years.

While it is very nice to have good neighbours, the pleasant thing about the farming community is that the neighbours flock to the farm sale to try to give the farmer and his family a good send-off when they have to sell up. Most of the neighbours will also buy something, whether they need it or not, just to help contribute. This is a long-standing tradition, and let us hope that it continues. Other people go along to buy things, but others only attend to be nosy as some farmers would say. Some of those attending may have had some connection with the farm or the farmer in the past, and they visit the sale for the sole purpose of meeting the farmer again.

The numbers of farm sales in some parts of the country will continue to increase as more and more farmers will have to leave the industry. This has been caused by some types of farming becoming less profitable, the production of food seeing more controls, and money continuing to be expensive to borrow.

A farm open day in Galloway, Scotland in the 1980's. They had belted Galloway cattle, Highland cattle, and a great variety of other farm animals to see.

Some old milking units at a farm sale, once used on every dairy farm.

Mr. Bill Johnson from Ponteland near Newcastle-upon-Tyne walking one of his dairy cows round the pen at his farm sale in 1981.

A farm sale in Northumberland in 1993. The sale attracted a big crowd of people. It was also a lovely day.

Hexham Cattle Market selling sheep.

A suckler cow and calf being sold at Hexham Cattle Market.

CHAPTER 13

MANAGING THE COUNTRYSIDE

Managing the countryside is just like managing a large garden with all different types of trees, shrubs, rivers, ponds, pathways, moors, hills, and all the various different types of land and soil.

FARMING AND WILDLIFE

There is nothing nicer than seeing a pair of Clydesdale horses ploughing a field, and the black-headed gulls hovering behind the plough looking for worms and mice on a winter's day. Although it is called a black-headed gull, the top of its head is really a chocolate brown colour. The black-headed gull is known as the farmer's friend because it can be seen following the tractor and plough or horse and plough furrow after furrow, hunting for mice and worms. These gulls are very sociable birds, and live in colonies. In the winter, the brown head feathers moult, and are replaced with white feathers. The black-headed gull is one of the smaller gulls in Europe, and is about fifteen inches (38 cm.) from head to tail.

To plough a straight furrow takes both art and skill. There are many ploughing matches held every year in various parts of the country for both tractors, horse, and for vintage tractors, all trying to plough a straight channel, which has to be a certain depth and width. Not as easy as it looks, but great fun!

All the hedges on the farm are normally cut and trimmed in the winter, and it is nice to see a tidy farm with all the hedges on the side of the road being well cut. At the turn of the year, about March, all the farm buildings and farmhouse had to be painted. If most of the buildings were built with stone, all the doors and cutters on the farm were painted. This is how many farmers in the south of Scotland operated, and it was known as spring cleaning. Also, in the spring of the year, all the farm fences were repaired before any cattle or calves were turned out to grass. Before any spring crops are sown, the winter ploughed fields will be worked with discs to break down the soil, then the land is harrowed, and the crop is sown. These spring sown crops could be new grass seeds, oats, or barley. The field was never rolled until the crop was up through the ground. The lapwing, or peewit as we know it better, would scrape a hollow nest in a field, and lay three or four eggs. When we came to roll the field, we had to look out for the peewits' nests. The ploughman and some of the other farm workers would sometimes pick up a few eggs, take them home, and have them fried. The peewit got its name because of its call; pee-wit, pee-wit. Over this last thirty or so years, their numbers have been declining, partly because of the changing methods of farming.

It is always a good time on the farm in the spring of the year, when the dairy cows and calves are turned out to grass. Milk yields shoot up in a few days, with the cows eating the new flush of grass. The young stock first turned out to grass run around the field, kicking their heels up in the air with excitement at their new freedom after being shut up all winter. There are more dairy cows down the west side of the country because of the higher rainfall. With it being wetter down the west coast, you get more insects such as worms and beetles.

There seem to be more cases of brucellosis cropping up on the farms among the dairy cows. The finger is pointed at the badger as carrying, and spreading the disease, which can be spread to humans. This is a very tricky question concerning whether the badgers spread brucellosis or not, and the experts are looking into this. What I would say is that the countryside would be a much poorer place to visit without the badger being around, as they are lovely sociable animals. There are trials going on at the moment about T.B. and badgers, and the results will be known next year, 2004. The badgers do mix among dairy cows and other farm animals, as they hunt at night for food, such as worms, beetles and berries. When a badger gets a tusk broken by a car, or with old age, it will come into the farm buildings looking for food such as cow nuts, milk powder, poultry food, or any other food lying around. Badgers are fairly big animals, and can be as big and heavy as labrador dogs. On average, badgers are about thirty six inches (91 cm.) long, that is from the tip of their nose to the tip of their tail. The hair from the badger was at one time made into shaving and paint brushes, and its grease was supposed to be an excellent cure for wounds. It is also thought that, at one time, the hams of the badgers were eaten in some areas.

The countryside is a lovely place to live and work. I can't see much change now in the countryside since I was a boy in the 1940's; though there are probably less trees, hedges, and stone walls now. Our family have always improved the parts of the countryside we own, both for the environment and for the wildlife over the years. Generations of our family have always been interested in the country sports of hunting, shooting, and fishing, which are getting a bit of a knocking at the present time. We, as a family, have always put more back into the countryside than we have taken from it, and have kept improving it for other people to enjoy. I keep striving to improve our farm by planting trees and shrubs, and creating bits of areas for the wildlife. I am very well aware of the feelings of some people and some groups who are against chasing foxes with dogs, shooting game birds and fishing. Let me tell you a bit about what our family do, then you can make up your own mind about us. First my brother, who works full time as a whipper-in with a hunt in England.

FOX HUNTING

When a younger brother of mine was a youth, he was very interested in horses. I can remember him buying a dark coloured rough haired fell pony. What a lovely creature this animal was! My brother was only in his teens at the time. The local hunt wasn't all that far away, and they used to hunt on a Wednesday and Saturday every week. The following Saturday after my brother bought the pony, he was off with the local hunt, and this is how he started fox hunting. They must have taken to him at the hunt as the hunt master invited my brother to come back to hunt with them.

After this, my brother spent all his spare time at the kennels helping to look after the fox hounds and the horses. He also used to go to the farms to collect the fallen stock with a trailer to feed the fox hounds. His future career was set, as this is what he decided he wanted to do. He worked his way up through the hunting ranks, and is now a whipper-in with a large well-known hunt in the Midlands.

My brother is not a cruel person, and he loves wildlife and the countryside. Fox hunting to him is a job, and he is there to control foxes in his area which kill the farmers' lambs, chickens, ducks and geese. He has experienced the pressure from some M.P.'s and some of the general public to have fox hunting banned, as they think it is cruel. They have a right to express their views.

In 1786, a group that were out fox hunting were sued for trespass at the King's Bench Court. The case was flung out. In recent years, judges have been careful never to commit themselves so far as to say that there is even any limited right of pursuing vermin which has been started on one man's land, and followed onto the property of another.

An older brother of mine, who passed away this year, was a great rod fisher man. He loved his trout fishing, and he used to tie his own flies at one time. There wasn't very much that he didn't know about trout fishing and the rivers. This brother also loved the countryside and the wildlife. He often used to tell me about some of the wild animals and birds that would nearly come up to him when he was sitting silently on the river bank fishing away. He used to say that the dipper, kingfisher, and various species of ducks would regularly also be fishing right beside him. While fishing on this well-known river, he sometimes saw an otter passing by, close into the river bank. He said that the otter was sometimes as much surprised as he was, as their eyes clashed. As soon as the animal caught sight of him, it stopped, and looked at him for a few seconds, then moved smartly on. My brother often brought trout home which the family enjoyed, but I think that he enjoyed watching the wild life from the river bank as much as he revelled in the fishing.

Then there is myself. I am the game shooter in the family, and have followed what my father enjoyed doing. He loved working his gun dogs in the countryside every bit as much as he enjoyed shooting the game. I have always had gun dogs, labradors and spaniels, and have enjoyed working with them, as well as game shooting, although I don't have any just at the moment because of my travel movements.

I love the countryside and everything that goes with it. Game shooting is not just about killing things, far from it. The enjoyment is also being out in the countryside with friends and their gun dogs, and the food and drink that is provided for us is always delicious. Most of my landowner and farmer friends that game shoot are always improving the countryside by planting trees and shrubs, and digging ponds which the wildlife in the countryside all benefit from. Therefore the general public can enjoy what the landowners and farmers do for them.

Things just don't happen in the countryside. As I have said, I keep planting trees, shrubs and flowers, which provide cover and shelter for wildlife, and they help to increase the beauty of the landscape for people to see and enjoy. I have a fairly large duck pond, and I continually feed the ducks and the other wild birds that visit the pond. Around the margins there are trees and shrubs, which make the place a lovely wildlife habitat. I shoot the ducks one or two nights in the winter with some friends. Then my wife Kathleen makes us a lovely summer pudding to finish off the evening. The small amount that I take from the duck pond does not compare with what I put back into the pond area. In these winter months I feed hundreds of birds such as sparrows, finches, crows, pheasants, partridges and many other species of wild birds that come there to look for food. Some people would say that feeding all these wild birds every winter doesn't give me the right to shoot the ducks, as shooting is cruel. I accept what some people will say, but I think the countryside would be a much poorer place to visit by the general public if it wasn't for country people like myself.

WOODLAND MANAGEMENT

When you visit the countryside you will see many single trees in fields and pastures, and in the hedgerows throughout the landscape. You will also see small and large plantations of various sizes. The trees, woods and plantations are not just growing wild. Many landowners and farmers have planted these for various reasons. There are also many single trees planted due to the fact that some people like certain specific trees, such as oak, beech and lime.

Many woods have been planted for sport and for shelter belts for cattle. A good sporting wood is one which is warm, with good nesting cover and roosting branches for the game birds. Woods and trees planted for shelter can be a great asset on a farm, especially where the buildings are on an exposed hillside, and where the livestock are kept outside all the year round, as the plantations protect both buildings and animals from high winds, lashing rain, and from a snow blizzard. The woods also help to conserve the wildlife in the area.

The fast growing trees, such as the larch, Scots pine, Douglas fir, and Sitka spruce make good shelter belts in the shortest period of time. However, some conifers, depending on the species, take between sixty and eighty years to mature. The trees have then reached an age where they can be felled and harvested. It is very important that if woods are to be effective at providing shelter, and for holding game birds and wildlife, then the woods should be designed properly.

Conservation means to protect the natural environment, and to look after the things around us, trying to protect the various aspects of the countryside and the buildings. The weather can have a great effect on many trees and woods. For instance, in 1978 millions of trees were uprooted and blown over by the hurricane winds which were devastating, especially in the south of the country where millions of pounds worth of damage was done to the trees and woods by the strong winds. Most recently, on Boxing night in 1998, millions of trees were again uprooted and blown over, especially in the north, by the strong winds, and a lot of damage was also done to houses, farm buildings as well as property the same night.

Here are some of the timbers which are grown in this country, and which enhance the countryside throughout the year with their different shades and colours.

HARDWOODS	SOFTWOODS
Oak	Scots Pine
Beech	Douglas Fir
Cherry	Norway Spruce
Ash	Larch
Walnut	Sitka Spruce
Sycamore	
Chestnut	
Poplar	

There are some hardwoods in this country which have been standing for hundreds of years, and have footpaths throughout the woods where the general public can walk and picnic amongst some of the oldest and most beautiful trees in the country. In the north east of England, for instance, there are about a quarter of a million acres (100,000 hectares) of woodland of various species, and in England as a whole, there are about five million acres (two million hectares) of woodlands.

***

A badger at the sett entrance. Note the huge amount of earth dug out by them.

A badger sow with her cub.

A young fox cub which is about ten weeks old.

A fox, having his leg repaired after being hit by a car.

The Haydon Hunt. Fox hounds up near the Roman Wall in Northumberland.

Neil Milbourn from Walby Grange, Crosby on Eden, having a five minutes break while out looking over his live stock with his dogs.

A cock pheasant.

Wenty Beaumont with one of his forestry workers, Graham Churnside, at Bywell near Stocksfield, preparing to have a tree felled.

Hardwood trees being planted by the Forestry Commission beside a stream in the Castle Douglas Forest in Galloway in the south west of Scotland.

Loading timber at the roadside.

The author in a plantation which has just been tidied up.

Landowners and forestry workers being presented with shields and trophies by the Forestry Commission at the 1986 Royal Highland Show in Edinburgh, for their contribution to forestry.

This is to certify that

Hunter Adair

was a runner-up in
the Countryperson Of The Year
category at the 2002
NFU Countryside Awards

Ian Dalzell
Chief Executive
NFU Countryside

Sponsored by

Also by the same author

MUCK SPREADIN'
SHOOTING AND THE COUNTRYSIDE
MUCKY BOOTS
THE MUCKY ROAD TO TONGUE FARM
THE FOUR SEASONS IN THE NORTH
SOME WILDLIFE SECRETS
THOMPSONS OF PRUDHOE:
THE FIRST FIFTY YEARS

CHILDREN'S BOOKS:

STONE WALLS
CLYDESDALE HORSES
FRUITS, BERRIES, FLOWERS AND SHRUBS
FARM ANIMALS AND FARMING
WILD ANIMALS AND BIRDS
HAY AND HARVEST TIME
FARM BUILDINGS AND MILKING PARLOURS
PEOPLE WHO LIVE AND WORK IN THE
COUNTRYSIDE